DUCK FARMING FOR

BEGINNER

I0477356

A Guide To Raising Healthy Pochard, Sustainable Practices, Egg Production, And Profitable Farming Techniques

Holden bodhi

Contents

DISCLAIMER

The information provided in this book, is intended for educational and informational purposes only. The content is based on research, personal experiences, and general knowledge about farming. It is not intended to substitute professional advice or expert consultation. Readers are encouraged to seek professional guidance when implementing any practices or techniques discussed in this book.

The author and publisher make no representations or warranties of any kind regarding the accuracy, applicability, or completeness of the contents of this book. Any reliance you place on such information is strictly at your own risk. The author and publisher shall not be held liable for any damages, losses, or injuries resulting from the use of the information provided.

Additionally, the author does not endorse, recommend, or affiliate with any individual, product, service, website, organization, or brand mentioned or referenced in this book. Any such references are solely for informational purposes, and no warranty or guarantee is implied. The inclusion of these references does not imply any endorsement or partnership by the author.

CHAPTER ONE

Overview Of Duck Farming

Around the world, duck farming has been a popular and lucrative enterprise for many centuries. If you take the time to learn about the ins and outs of duck farming, you can reap many benefits from raising ducks for eggs, meat, or just as a pastime. Ducks are a good choice for novices since, in contrast to chickens, they are more robust to bad weather and require less care.

It's important to take into account several things before beginning your duck farming endeavor, including the breed of duck you wish to raise, the kind of housing needed, the feeding schedule, and basic maintenance. The scale of duck farming can be adjusted to suit your needs, from a tiny backyard operation to a large-scale farm. In addition, ducks can help your farm or garden's soil quality and are great at controlling pests.

We'll go over the key components of duck farming for novices in this tutorial, giving you the information you need to start a profitable farm and enjoy the rewards of this fulfilling hobby.

Why Raise Ducks For Farming?

Although they are sometimes disregarded as farm animals, ducks are a good starting animal for new farmers. Ducks are resilient animals that can adapt to a variety of climates and situations, unlike hens. They are a more manageable option for beginning farmers because they are less susceptible to many of the diseases that strike other birds.

The adaptability of ducks is a major factor in the popularity of duck farming. There are several reasons to raise ducks, including:

• Egg Production: Duck eggs are more nutritious, bigger, and frequently sell for more money than chicken eggs. They are particularly well-liked in the baking sector because of their richness and capacity to improve baked items' texture.

• Meat Production: Duck meat is prized in many cultures throughout the world. It is a sought-after ingredient in gourmet cooking since it is more flavourful and contains more fat than chicken.

• Feathers and Down: The textile industry places a high value on duck feathers and down. Because of their insulating qualities, they are frequently used to make cushions, blankets, and clothing items.

• Pest Control: Being omnivores by nature, ducks like consuming pests like slugs, snails, and other insects that pose a threat to crops. They are therefore an environmentally beneficial pest management option for organic farming.

Ducks also require less upkeep than other animals. They can supplement their diet with natural foraging, and they don't require sophisticated housing. Ducks present a low-cost, high-reward potential for beginning farmers seeking a manageable entry point into animal farming.

Advantages Of Duck Raising

There are several advantages to raising ducks, for the farmer as well as the environment. You may start duck farming with more informed decisions if you are aware of these advantages.

1. Resistance to Disease and Hardiness

Ducks are renowned for their capacity to withstand and adjust to a wide range of weather conditions, including chilly winters and scorching summers. Ducks require less care than chickens because they are less prone to several common poultry diseases. For new farmers who might not be acquainted with the complexities of disease prevention and treatment in chicken farming, this is especially crucial.

2. Durable Termite Management

Raising ducks has several important environmental benefits, one of which is their natural capacity to control pests. Ducks are great foragers; they will gladly eat slugs, insects, and other minor pests from your farm or garden. This not only lowers the requirement for chemical pesticides but also enhances soil quality because duck droppings act as a natural fertilizer.

3. Profitable Production of Meat and Eggs

Meat and eggs from ducks are premium goods. Duck eggs are increasingly popular in niche markets, particularly among bakers and health-conscious consumers, because they are bigger, richer, and packed with more nutrients than chicken eggs. Duck meat, especially from Pekin and Muscovy breeds, is highly valued for its flavor and texture and can fetch high prices in gourmet and local markets.

4. Minimal Upkeep Needs

When compared to other animals, ducks take very little maintenance. They don't require as much housing as chickens do, and they may be kept in a range of conditions. Ducks don't need complex nesting boxes or tall fences, and their tendency towards aggression is lower. Your ducks will be healthy and happy with just a basic, safe home and access to clean water.

5. Pets and Companion Animals

Ducks are not only beneficial economically, but they may also make excellent friends. They are an excellent option for families, especially those with young children, as they are typically amiable and less combative than hens. Because they are intelligent animals, ducks can even be taught to obey simple commands, which makes caring for them enjoyable.

An Overview Of Basic Duck Farming

Knowing the fundamentals of duck farming is essential before establishing a farm. Duck farming involves several essential elements, such as choosing the appropriate breed, building a suitable home, providing food for the ducks, and preserving their well-being.

1. Selecting the Appropriate Breed of Ducks

There are many different breeds of ducks, and each has certain qualities and advantages of its own. While certain breeds are valued more for their meat, others are better at producing eggs.

Novices must choose a breed that complements their farming objectives. Some popular breeds are:

• Pekin Ducks: Distinguished by their quick development and high meat yield.

• Khaki Campbell Ducks: Known for producing an abundance of eggs.

• Muscovy Ducks: Highly valued for their ability to control pests and provide lean meat.

2. Needs for Housing and Shelter

It's not too difficult to house ducks. As long as they are shielded from predators, they may survive in a variety of shelters and don't require roosting bars like chickens do. It's critical to give ducks access to water for swimming and drinking, as well as a clean, dry place to rest.

Ducks require water to maintain good feather health in addition to staying hydrated.

3. Nutrition and Feeding

Being omnivores, ducks may survive on a diet consisting of grains, insects, and water plants. Egg production and good growth depend on a balanced diet. Either buy duck feed or let your ducks forage for themselves, adding grains and leftover food from your kitchen to their diet. Ducks need to submerge their heads in water to clear their noses and preserve the health of their feathers, thus fresh water should constantly be provided.

4. Management of Health and Disease

Even though ducks are often resilient, it's crucial to keep an eye out for any indications of illness or discomfort. Maintaining your ducks' health primarily involves giving them fresh water regularly, giving them a balanced meal, and cleaning their living quarters. Depending on your location and the number of your flock, vaccinations, and worming may be required.

Establishing a duck farm can be difficult initially, but with the correct planning and knowledge, it can be a very fruitful endeavor. You may build a profitable and sustainable duck farming business that helps the environment and yourself by understanding the principles of the industry.

CHAPTER TWO

Selecting The Appropriate Breed Of Ducks

Choosing the proper breed is an important first step in duck farming. Ducks come in a variety of breeds, each with unique traits, advantages, and disadvantages. The breed you choose will be determined by your particular objectives, such as whether you want to raise ducks for meat, eggs, or decoration, or all three. Starting with the correct breed at the outset can make all the difference for new farmers in terms of a successful and trouble-free experience or one full of needless obstacles.

The Reason For Breeding Ducks

Finding the reason for raising ducks is the first step in selecting a breed. Which is your main interest: producing meat, eggs, or both? Some breeds are better suited for particular tasks. For instance, some breeds grow more quickly and are more suited for producing meat, while others make outstanding egg layers. There are also ornamental breeds that are raised mainly for their distinct look and peaceful disposition, which makes them excellent pets or display animals.

Before you even start looking for the perfect breed, it's important to know what you want from the breed. Given that ducks are known to consume slugs, snails, and other insects that can harm crops, some individuals may also breed ducks as a hobby or as a means of controlling pests. Your selections will become more limited if you have a defined aim in mind.

Ducks Laying Eggs

Indian Runner and Khaki Campbell are great breeds to raise if your main objective is to produce eggs from ducks. These ducks are renowned for laying a lot of eggs. For example, one of the most popular varieties among egg growers is the Khaki Campbell, which can lay up to 300 eggs annually. Another popular breed for eggs is Indian Runners, who are renowned for their erect gait and exuberant disposition. Each year, they can yield 200–250 eggs.

Ducks That Produce Meat

The greatest options for farmers who are focused on producing meat are the Pekin and Muscovy ducks. Because of their quick growth and superb meat quality, Pekin ducks are a favorite in the commercial duck meat sector. Their flesh is flavourful and soft, and they can be ready for the market in as little as 7-9 weeks. Conversely, the Muscovy is a big breed valued for its flavourful, lean flesh that is frequently likened to veal. They are a desirable alternative for people living in more suburban or residential environments because they are quieter than other kinds.

Breeds with Two Uses

Breeds such as the Rouen or the Welsh Harlequin are good choices if you're searching for ducks that can lay both eggs and meat. The quantity of meat and eggs is balanced in these ducks. The Welsh Harlequin gets to a respectable size for producing meat, but it is also a great egg layer, laying 200–250 eggs a year. Despite growing more slowly than Pekins, Rouen ducks are a good dual-purpose duck since they are larger and resemble Mallards.

Popular Breeds Of Ducks For Novices

It can be difficult for a novice to choose the correct breed, but several breeds are well-known for being docile and having a variety of uses. Some of the most well-liked duck breeds for novices are listed below:

Campbell, Khaki

With 250–300 eggs per annually, the Khaki Campbell duck breed is among the most widely used for egg production. They are easy to handle, even for novices, because they are little, energetic ducks with a placid demeanor. They also lay big, beautiful, and high-quality eggs. Although Khaki Campbells are not usually grown for meat, they are ideal if producing eggs is your main objective.

Pekin

The most popular breed used to produce meat is the Pekin, which is renowned for its enormous size and quick growth. In as little as seven weeks, they can be processed, and they are amiable and simple to raise. Because there is less obvious coloring remaining on the carcass, processing is also made easier by their white feathers. Pekins may lay a respectable quantity of eggs, usually between 150 and 200 per year, despite being primarily grown for their meat.

Runner in India

Indian Runners are a distinct breed distinguished by their high levels of energy and erect stance. They lay 200–250 eggs a year, making them exceptional egg layers. Because of their low weight, these ducks are perfect for farmers who have little room. Moreover, Indian Runners are renowned for their capacity for foraging, which can lower feed expenses.

Moscow Muscovy

Larger and more subdued, Muscovy ducks are prized for their flavourful, lean flesh. They can be a little territorial and are a little more independent than other breeds, but they are great at controlling pests and need less feed than other meat ducks. Although it takes them 12 weeks to reach market size, muscovies are known for their sluggish maturation and highly sought-after flesh.

Considerations (Climate, Purpose)

The environment, the goal of your farm, and the available area are all key considerations when selecting the best breed of ducks. These all have an impact on which breed will do best in your specific farming situation.

Temperature

Certain duck breeds are better adapted to warm regions than others, while some are more resilient than others in the face of extreme cold. Pekin ducks, for example, are comparatively resilient and adaptable to many conditions, although in colder temperatures they will require appropriate housing. Conversely, Indian Runners thrive in warmer regions but may suffer in cold, damp weather. Seek for breeds like the Muscovy, which are more tolerant of the cold, if you live in an area with severe winters.

Why We Farm

The goal of your duck farming will affect the breed you choose, as was previously mentioned. Go for prolific egg layers like the Khaki Campbell if your goal is to generate eggs. Pekin or Muscovy ducks would be more suited for meat production. Both the Welsh Harlequin and the Rouen are great dual-purpose breeds for people looking for a balance between meat and egg production.

Area and Nutritional Needs

While some breeds do better in smaller settings, others need more room and are more suited to larger farms. For example, Khaki Campbells are happy in tiny, cramped areas, but Indian Runners require a lot of room to roam around as they are active foragers. Furthermore, some breeds—like the Muscovy— can obtain a large portion of their nutrition by foraging, which lowers feed costs, while other breeds—like the Pekins—need more extra feed because of their quick growth.

How To Find Well-Being Ducks

Finding healthy ducks is the next step after selecting the breed that best suits your requirements. The success of your farm depends on making sure your ducks come from a reliable source. When it comes to finding ducks, hatcheries, breeders, and nearby farms are some of the alternatives.

Hatcheries

Purchasing ducks from a commercial hatchery is one of the most popular methods. Ducklings that are one day old can be supplied straight to your farm by many hatcheries. This might be a practical choice because hatcheries usually have a large selection of breeds available, which is especially useful for beginners. Seek to choose hatcheries with a solid track record of producing disease-free, healthy birds. It's crucial to find out if they vaccinate their birds or offer health certificates.

Breeders

Purchasing ducks from a nearby breeder can be a great choice if you're seeking a particular breed or want more individualized guidance. Breeders can give you comprehensive details regarding the temperament, health, and capacity to lay eggs or produce meat of the breed. They also frequently take great care in raising their ducks. Make sure to enquire about the immunization history and health history of the breeding stock when buying from a breeder.

nearby farms

You might also purchase ducks from a nearby farm. In addition to making sure the ducks are acclimated to the local climate, this is a terrific way to support local agriculture. Checking out the ducks in person at a nearby farm will reassure you about their well-being and living circumstances.

Considering Health

It is vital to seek indicators of good health while choosing ducks. Healthy ducks should be lively and aware, with clear, glossy feathers and sparkling eyes. Ducks who seem listless, have drooping wings, or exhibit respiratory symptoms like coughing or wheezing should be avoided. To stop the spread of disease, always quarantine new ducks for a few weeks before bringing them into your current flock.

CHAPTER THREE

The sections in your book titled Duck Farming for Beginners that deal with "Setting Up the Duck Housing" have a detailed overview available here. About 500 words are allotted to each segment, which is detailed with subheadings and information arranged into multiple paragraphs.

Putting The Duck Housing In Place

The health, happiness, and production of your ducks depend on you providing them with an appropriate living space. Ducks need specialized housing that offers them comfort and room to roam about while shielding them from predators and harsh weather. The fundamentals of establishing efficient duck housing will be covered in this section.

Ducks' Requirements For Housing
Density and Space

Ducks are gregarious animals that prefer living in flocks. For this reason, having enough space is essential to their mental and physical health.

A minimum of 4 to 6 square feet of indoor area and an extra 10 to 15 square feet of outdoor space should be allocated to each duck. In addition to causing tension and hostility, overcrowding can make people more susceptible to illness.

Safety

Ducks require a secure and safe habitat. Predator-proof housing is necessary to shield inhabitants from foxes, hawks, and raccoons. It's essential to have sturdy walls, fences, and locks. Hardware cloth is tougher and more resilient to predator assaults, so it is advised to use it instead of chicken wire.

Mattresses

In order to preserve comfort and hygiene in the housing, proper bedding is necessary. You can use straw, wood shavings, or straw pellets as bedding. These materials are not only cozy, but they also take in smells and dampness. Cleaning and replacing bedding regularly helps stop the growth of dangerous microorganisms.

Ground conditions and drainage

In order to avoid waterlogging in the duck housing area, proper drainage is essential. Due to their untidy nature, ducks have a tendency to transform the ground into muck, which can cause illnesses and foot issues. Sloped terrain can help ensure appropriate drainage and elevated duck homes can help lessen the amount of mud surrounding the living area.

Duck Shelter Types

Transient Shelters

Temporary shelters are a wonderful alternative for novices, particularly if you're starting with a small flock of ducks. You can build temporary enclosures or portable coops out of materials like wood, tarps, or even old pallets. These structures are mobile, allowing for the provision of new grazing spaces while limiting overgrazing in one location.

Long-Term Accommodation

Purchasing permanent housing becomes crucial as your duck farming business expands. A well-constructed duck home should be roomy and safe. During severe weather, insulated walls can assist keep a pleasant temperature. If you want to build your home with windows to let in natural light, be sure they can be closed off in inclement weather.

Run Outside

For ducks to get exercise and enjoy the sun, they need to go for outdoor runs. It ought to be big enough for ducks to walk around freely and have sturdy fencing. Incorporating organic features like trees or bushes can also offer cover and shade.

Water-related Features

Ducks adore water and depend on it for bathing as well as drinking. Ducks are content when there is a tiny pond or kiddie pool in the housing area since it promotes natural behaviors. To preserve hygiene, make sure that water sources are cleaned regularly.

The Value Of Airflow And Room

Airflow

Having enough ventilation in duck housing is essential. Ducks expel a lot of moisture and waste, which if left unchecked can cause respiratory problems. Ensuring proper ventilation can be facilitated by installing exhaust fans, vents, or windows. In hot temperatures, proper ventilation helps control temperature and lessens heat stress.

Spatial Considerations

Ducks require room to move about and engage in their natural activities, which include swimming, preening, and foraging. Aggression, tension, and health issues can result from crowded settings. A proper space reduces fighting among ducks and enables them to form a social hierarchy.

Temperature Regulation

The housing may need extra elements for temperature management, depending on the local region. Ducks can be kept warm in cold climates with insulation and heat lamps and cooled in hot climates with misters and shade structures. To keep things healthy, keep an eye on the duck house's humidity and temperature at all times.

Mental Health

Giving ducks enough room and a well-ventilated habitat is beneficial to their general behavioural health. Ducks that are allowed to roam freely and engage in their natural activities are less likely to experience stress-related ailments. A happier, healthier flock is encouraged by this conducive atmosphere that fosters natural connections.

In summary

Duck housing must be carefully set up with respect for ventilation, security, space, and surrounding surroundings. Knowing these needs increases the likelihood that your ducks will flourish, which will eventually result in a fruitful and satisfying time raising ducks.

CHAPTER FOUR

Nutrition And Feeding

Nutrition and feeding are essential components of a profitable duck farm. A well-balanced diet is necessary for ducks to be healthy, lay high-quality eggs, and grow as quickly as possible. You can give your ducks the finest care possible if you are aware of their dietary requirements.

The Basic Diet Of Ducks

Because they are omnivores, ducks eat both plant and animal materials. Grain-based and vegetable-based diets, as well as protein sources and vital minerals and vitamins, should be well-balanced.

Corn, wheat, and barley are examples of grains that are essential to a duck's diet. The majority of their diet ought to consist of these grains, which give them energy. About 16–18% of a duck's diet must consist of protein, which is essential for growth, egg production, and general health.

Commercial duck feed that is made especially for ducks and offers a balanced combination of grains, protein, and vital elements is one source of protein. If you would rather make your own feed, you could want to add fish meal, soybean meal, or peas as protein supplements.

Greens and vegetables are also necessary. Greens that are high in vitamins and minerals, such as spinach, kale, and lettuce, are popular among ducks. Providing kitchen scraps such as fruits, cereals, and vegetable peels can be a great way to add variety to their diet and cut down on waste.

When creating a diet plan, keep in mind that introducing new foods gradually can help avoid digestive problems. To keep them healthy and productive, keep an eye on how they react to dietary changes and make the necessary adjustments.

Ducks' Requirements For Water

Water is essential for a duck's wellbeing. Ducks have different water needs than chickens since they use water for bathing as well as drinking.

There should always be clean, accessible drinking water. Fresh water is a must for ducks since it facilitates their digestion and helps them absorb nutrients from their meal. Having enough water available is crucial, particularly if they are eating dry food.

Water for bathing is still another crucial factor. Ducks maintain their cleanliness and well-being by using water to clean their feathers, eyes, and nostrils. Give the ducks access to a big water trough or a little pond so they may bathe and dip their heads. This maintains their feathers healthy and helps avoid sickness.

Make sure there is an ample quantity of water available during hot weather since ducks can quickly become dehydrated. Always make routine checks on water sources to avoid pollution. If you use a water trough, clean it often to prevent toxins and algae buildup that could hurt your ducks.

Feeding Plans And Additives

You must set up a feeding regimen if you want to keep your ducks healthy and productive. Generally speaking, ducks need two to three meals a day, depending on their age and stage of production.

Ducklings, or young ducks, should be provided starter feed that is appropriate for their stage of growth. To guarantee that they are receiving the nutrients required for healthy growth, feed them at least three times a day. Once they reach adulthood, you can switch them over to grower feed, which you can give them twice a day.

The number of times an adult duck can be fed each day will vary based on how productive they are. Layer feed, which has greater calcium levels necessary for robust eggshells, should be available to layer ducks reared for their eggs. Maintaining a regular feeding schedule guarantees that they get enough nutrition all day long and aids in digesting control.

Your duck's nutrition may also be significantly impacted by supplements. Their health can be improved by supplementing their diet with vitamins and minerals, particularly during stressful periods like molting or right after immunization. Goods such as calcium supplements for layer ducks can support general health and egg quality maintenance.

Give your ducks some grit; it helps them break down food in their gizzards, which helps with digestion. Make sure the grit is tiny enough for the ducks to easily eat it, and top it off as needed.

To sum up, offering a healthy diet, hygienic water, and a consistent feeding schedule are essential elements of duck farming. Knowing what your ducks need to eat will help you raise successful ducks and make sure they thrive.

CHAPTER FIVE

Duck Well-Being And Illness Avoidance

Even if it's a fulfilling job, raising ducks has its difficulties, especially when it comes to the flock's health. For both novice and experienced duck farmers, it is essential to comprehend the significance of health and disease prevention. Common duck diseases, vaccination schedules, biosecurity protocols, and identifying symptoms in your ducks are all covered in this section.

The Significance Of Health Administration

It's critical to your ducks' general health and production to keep them in good health. Healthy ducks grow more quickly, lay more eggs, and are more resilient to illness. Proactive health management includes regular check-ups, healthy eating, and knowledge of typical health problems that may occur. Establishing a thorough health management program makes farming more successful by reducing the likelihood of disease outbreaks.

Environment and Nutrition

Your ducks' health is mostly dependent on their nutrition being balanced. The nutritional needs of ducks differ based on their age, breed, and intended use (producing meat or eggs). Make sure the food your ducks eat is nutritious and full of grains, proteins, vitamins, and minerals. Additionally, a duck's living habitat is very important to its health. Provide hygienic, dry housing with sufficient ventilation to avoid respiratory problems and moisture-related illnesses.

Typical Duck Illnesses

Ducks are susceptible to a variety of diseases that can affect their health and productivity, just like any other livestock. It is essential to familiarise oneself with these illnesses to manage and avoid them effectively.

1. Bird Flu

Ducks can contract avian influenza, also referred to as bird flu, which is a viral infection. It can vary in severity, with certain strains posing serious risks to the health of humans and ducks. In severe cases, symptoms could include decreased egg production, respiratory difficulty, and sudden death. Preventive actions include keeping oneself biosecure and avoiding wild bird interaction.

2. Duck Hepatitis Viral (DVH)

A virus that affects young ducklings is the source of the highly contagious disease known as duck viral hepatitis. Signs of lethargy, enlarged liver, and significant mortality rates are possible in infected birds. Assuring proper biosecurity procedures and immunizing ducklings at a young age, as advised by a veterinarian, are the best preventive measures.

3. Botulism

A toxin generated by the bacteria Clostridium botulinum, which is frequently present in decomposing organic waste, is what causes botulism. Ducks can contract an infection if they consume tainted feed or water. Weakness, paralysis, and incoordination are among the symptoms. Keeping the ducks' surroundings clean and making sure they have access to food and fresh water are essential to preventing botulism.

4. coccidiosis

Ducks, especially young ones, are susceptible to a parasite condition called coccidiosis, which is brought on by Eimeria species. Poor growth, weight loss, and diarrhoea are among the symptoms. If necessary, medicated feed can be given in order to prevent this sickness. Proper sanitation is also important.

Measures For Biosecurity And Vaccination

Vaccination schedules and biosecurity regulations must be followed to keep your flock of ducks disease-free.

Guidelines for Vaccinations

A vital component of illness prevention is vaccination. Speak with a vet to create a vaccination program that works for your particular farm. Newcastle disease, avian influenza, and duck viral hepatitis are among the common vaccinations. Vaccinating your ducks at the right ages and intervals can help them develop immunity and lower the likelihood of illness outbreaks.

Biosecurity Procedures

The term "biosecurity" describes the precautions used to stop the entry and spread of illnesses. Among the biosecurity techniques that work well are:

• Isolation of New Ducks: To check for symptoms of sickness, confine new ducks for at least 30 days before bringing them into your flock.

• Managed Access: Provide only necessary staff members access to your waterfowl area. At entry points, use footbaths with disinfection.

• Cleanliness: Keep feeding spaces, housing, and equipment clean by routinely cleaning and disinfecting them.

• Wildlife Control: Keep your duck area fenced in and stay away from open water sources to avoid coming into contact with scavenging birds and other wildlife that could be disease carriers.

Recognising Symptoms Of Illness

Promptly identifying symptoms of sickness in your ducks is essential for both their well-being and the general prosperity of your farm.

Modifications in Behaviour

A behavior shift in ducks is one of the first signs of disease. In general, healthy ducks are gregarious, inquisitive, and active. Ducks that appear lethargic, withdraw, or refuse food or liquids could be suffering from a medical condition. Keep a regular eye on their relationships and general levels of activity.

Symptoms in the body

Seek out any physical indications of disease, such as:

• Respiratory Problems: Nasal discharge, sneezing, or coughing may be signs of respiratory infections.

• Digestive Problems: Abdominal swelling, bloating, or diarrhea may indicate gastrointestinal problems.

• Lameness: Leg or foot injuries or infections may be indicated by limping or trouble walking.

• Feather Condition: Deficient nutrition or health issues may be indicated by feathers that are ruffled or absent.

Frequent Health Examinations

Regular health examinations are crucial for the early diagnosis of disease. Examine the weight, color of the comb, and general look of each duck. Maintaining a health journal might be useful for monitoring changes over time and seeing trends that might point to illness outbreaks.

In summary

You may drastically lower the danger of illness by making your ducks' health a top priority and taking appropriate biosecurity, vaccination, and monitoring procedures. Knowing the signs and symptoms of common duck diseases will help you take prompt, decisive action to maintain a healthy and successful duck farm.

CHAPTER SIX

Breeding Ducks And Producing Eggs

When it comes to breeding and egg production, duck farming may be a lucrative endeavor. Because of their adaptability and tenacity, ducks are a great option for beginning farmers. We'll look at duck reproduction in this section, including how they lay eggs when they lay them, and useful incubation techniques.

How Ducks Breed

Comprehending the reproduction process of ducks is essential for effective breeding. Ducks reach sexual maturity between the ages of five and six months, however, breed-specific differences may apply. The physical characteristics that set male ducks, often known as drakes, apart from females include brighter coloring and a distinctive quack.

Courtship Conduct:

Ducks have distinctive courtship rituals that differ depending on the breed. Males usually use vocalizations and physical demonstrations to assert their dominance. To get a female's attention, a drake might, for example, flap his wings, bob his head, or even do a peculiar "dance." It's crucial to watch these actions to determine when ducks are ready to mate.

Mating Procedure:

Mating happens when a female expresses interest; this usually happens quickly. Fertilization may result after a successful mating, in which case the male crawls atop the female's back. Maintaining an ideal male-to-female ratio—typically one male for every four to six females—is crucial for fostering ideal reproductive conditions and avoiding stress in the females.

Effective egg management requires an awareness of the cycles in which ducks normally lay their eggs. The majority of duck breeds lay between 100 and 300 eggs a year, depending on the breed, age, and surroundings.

Laying Structures:

As the days become longer in the spring, the ducks produce more hormones, which triggers the egg-laying cycle. Usually in the morning, a female duck can deposit eggs virtually every day. Frequent egg collection is necessary since letting eggs remain in the nesting location for extended periods might cause problems like pecking or cracked eggs.

Sites for Nesting:

Encouraging suitable nesting locations is essential. For laying eggs, ducks choose isolated, calm, and dark locations. To ensure comfort, an appropriate nesting box should be dry, clean, and packed with hay or straw. To encourage good laying and lower the risk of infection, check and clean the nesting boxes regularly.

Maintaining Records:

Keeping thorough records of your flock's egg production is essential to monitoring its well-being and productivity. Take note of the quantity laid, the quality of the eggs, and any variations in the pattern of laying. Your ducks' health problems and environmental stressors can be found using this data.

Duck Egg Incubation

A crucial first step in growing your flock is ensuring that your duck eggs are successfully incubated. For the best chance of hatching, duck eggs need a certain combination of temperature and humidity.

Organic Fertilisation:

Given a secure environment, ducks are known to be great moms and will naturally incubate their eggs. A broody hen will spend about 28 days sitting on her eggs. She will turn the eggs frequently throughout this period to guarantee uniform heating.

Synthetic Fertilisation:

Specialized incubators are available to mimic natural settings for individuals who prefer artificial incubation. A temperature of 99.5°F (37.5°C) and a humidity of roughly 55–60% should be maintained in the incubator. Turning the eggs several times a day—by hand or with an automated turner—is crucial.

Tracking Development:

One method for monitoring the development of the embryos inside the eggs is candling. You may see the growth and find out if the eggs are viable by shining a bright light through the egg's shell. Around the seventh and fourteenth day of incubation, respectively, this should be completed.

Getting Ready for the Hatch:

Raise the incubator's humidity to roughly 70% as the hatching date draws near to help the ducklings burst through their shells more easily. The ducklings should start to pip or break through the shell, on day 28. To prevent anxiety during this period, make sure the surroundings are calm.

In conclusion, anyone considering a career in duck farming needs to have a solid understanding of egg production and breeding. You may create a profitable duck farm by concentrating on appropriate incubation methods, efficient egg management, and appropriate reproductive practices.

CHAPTER SEVEN

Growing Up Ducklings

Growing ducklings may be a fulfilling endeavor for small-scale farmers as well as enthusiasts. Although ducklings are tough and durable in general, they need extra care and attention, especially in the beginning. To ensure the health and growth of ducklings, it is essential to comprehend their basic demands.

The Value Of The Early Years

A duckling's development is greatly aided by the initial few weeks of its existence. Taking good care of them at this time can have a big impact on their productivity and long-term health. For a duckling to flourish, they require warmth, food, and a secure home. Since their immune systems are still growing, they are more susceptible to illness. It is crucial to concentrate on their care in these initial phases.

Sombre Ambience

Raising healthy ducklings begins with creating a suitable brooding environment. Ducklings can grow in a warm, secure environment in a brooder. Bedding material, a heat lamp, and ample room for the ducklings to roam around are all common components of a brooding arrangement.

Heat Source: During their first week of life, ducklings need a warm atmosphere. Once they are fully feathered, the temperature can be progressively lowered by 5°F each week, starting at about 90°F (32°C). This temperature can be preserved with the use of a heat lamp hanging over the brooder.

Bedding: Use straw, wood shavings, shredded paper, or any other clean, dry bedding material. Retaining moisture should be avoided when choosing materials for bedding because it might cause health problems. Make sure the bedding is sufficiently thick to offer comfort and insulation.

Space Needed: Ducklings require room to move around and interact with others. It is advised to use a brooder that has at least one square foot for each duckling in order to reduce stress and crowding. They will be able to display their natural behaviours in this area, which is crucial for their growth.

Taking Good Care Of Ducklings

Ducklings require proper care to ensure their growth and well-being. A few essential components of care are keeping an eye on their health, making sure they're clean, and creating the ideal atmosphere.

Health Surveillance

Frequent health examinations are crucial for spotting any possible problems early on. Ducklings should have clear, bright eyes, be lively, and be inquisitive. Illness symptoms can include odd behaviour, lack of appetite, and fatigue.

It is imperative to seek the advice of a veterinarian with experience treating chickens if any health problems develop. Early intervention can stop minor difficulties from becoming more serious health problems.

Practices of Hygiene

It is crucial to keep the brooding space clean in order to avert disease outbreaks. Provide the bedding a regular cleaning and change, and provide the kid's food and fresh drink every day. Eliminate any rancid food to stop the growth of mold and lessen the possibility of luring bugs.

Furthermore, make sure there is adequate ventilation in the brooding area to lessen the accumulation of ammonia from waste. Ducklings may experience respiratory problems due to ammonia. Because ducklings like to splash, you should regularly check the water source for contamination.

Social Communication

Being gregarious animals, ducklings gain from social interaction within their group. Keeping kids in groups can lessen stress and aid in the development of social skills. To create a vibrant and busy atmosphere, think about adding a few additional ducklings if you're raising a small batch.

Providing Ducklings With Food

One of the most important components of caring for ducklings is feeding them correctly. They will get the nutrients they need for normal growth and development if they eat a balanced diet.

Initial Feed

High-quality beginning feed designed especially for ducks should be offered to ducklings. These diets typically provide important vitamins and minerals along with a high protein content of 20–24 percent. From the beginning, begin giving them this particular formula.

Pellets or Crumbles: Two forms of starter feed are available: pellets and crumbles. Ducklings can consume pellets more easily than crumbles, although crumbles might be less wasteful. Select a form based on the feeding preferences of your flock.

Water Availability: Ensure that there is always access to clean, fresh water. Since ducklings like to slobber during feeding, make sure their water source is both deep enough for them to wipe their nostrils and shallow enough to prevent drowning.

Treats and Supplements

You can add certain snacks and nutrients to ducklings' diets as they become bigger. Present modest portions of finely chopped fruits, grains, and greens. They still require the full nutrition that comes from their starting feed, so make sure these additions don't make up more than 10% of their daily intake.

Observing Feeding Patterns

Observe closely how your ducklings consume their food. They should routinely eat food and drink water. If any ducklings appear to be resisting food or water, this may be a sign of a health problem, therefore you should consult a veterinarian right once.

Establishing A Secure Atmosphere For Juvenile Ducks

For the sake of your ducklings' growth and well-being, you must provide a safe habitat. They are shielded from potential health concerns, predators, and environmental dangers by a secure space.

Setup of Enclosure

Take into account elements like size, security, and comfort in the surrounding surroundings when building the enclosure. Ducklings require a large amount of space for mobility, but the place must also be safe from predators.

Fencing: To keep predators away, choose strong fencing material that is at least three to four feet high. To keep animals from burrowing behind the fencing, make sure it is buried a few inches below the surface.

Shelter: Make sure ducklings have a place to go when bad weather strikes. Rain and direct sunshine can be shielded from them by a basic roofed building. Make sure the shelter is free of dangerous objects and sharp edges.

Enhancement of Environment

For natural development, ducklings require mental stimulation. Toys, mirrors, or other impediments can be introduced to foster playfulness in them. They get less stressed and improve both mentally and physically from this enrichment.

Safe Access to the Outdoors

When your ducklings are old enough, you can let them spend time outside under supervision. Make sure there are no potential hazards nearby, including poisonous plants or sharp items. Keep an eye on their actions to make sure they're having fun in the great outdoors safely and securely.

CHAPTER EIGHT

Taking Care Of Pond And Water Needs

Among poultry, ducks are special since they love the water. Water is vital to their health and wellbeing, whether they are swimming in a vast pond or dabbing in a small kiddie pool. This section will discuss the importance of water to ducks, how to build the perfect duck pond, and safe maintenance techniques.

Water Is Essential For Ducks

Ducks depend on water for several reasons. It facilitates their digestion first and foremost. Due to their peculiar digestive systems, ducks must drink water continuously for their bodies to properly metabolize food. Ducks must drink water while they eat, in contrast to hens, which can consume dry feed. Providing them with enough water keeps their digestive systems healthy and facilitates their efficient absorption of nutrients.

Water is essential for digestion as well as for keeping ducks' feathers healthy. Although ducks' feathers are waterproof, they nevertheless need to preen frequently to keep them in good condition. Ducks that have access to water can bathe, which helps them clean their feathers of excess oil, filth, and parasites. In addition to maintaining their personal cleanliness, preening helps them maintain body temperature since clean feathers act as insulation and buoyancy.

Water also facilitates social connections between ducks. Being gregarious creatures, ducks use the water as a gathering place where they may play, socialize, and form their social hierarchy. Their mental health depends on this relationship since it lowers stress and promotes natural behaviors. The general well-being of your ducks can be greatly improved with a clean, well-maintained water source.

Establishing a Duck Pond

Building a good duck pond is a worthwhile project that will help your flock tremendously. The following are important factors in setting up the perfect pond environment:

Place and Dimensions

Make sure your duck pond is in a spot that gets plenty of sunlight, shade, and accessibility. Ducks benefit from being in a sunny environment because it provides warmth and stimulates the growth of algae, which acts as a natural food source. But in warmer weather, it's crucial to have shady spots where ducks may hide.

The pond's dimensions are similarly significant. Ducks do well in tiny waterways, but larger ponds provide them greater room to swim, dive, and forage. In certain places, ponds should be at least three feet deep to enable ducks to dive and evade any predators. Make sure there are also gradual entry or mild slopes for easy access.

Water Purity

A duck pond's water quality is very important. Being untidy creatures, ducks can easily contaminate their water source with excrement and leftover food. Consider adding a filtration system or using a natural filtering technique, such as planting aquatic plants that can help absorb excess nutrients, to preserve the quality of your water.

It is crucial to regularly check the water's clarity and odor. Water that is clear suggests a healthy environment; water that is murky or has an unpleasant odor could suggest an issue. To avoid stagnation and algae accumulation, change the water in the pond regularly and clean the pond's margins.

Organic environment

Enhancing the ducks' experience in the pond is creating a natural habitat. To offer cover and nourishment, introduce water plants like duckweed, lily pads, and cattails. These plants support a healthy environment in the pond in addition to providing hiding places from predators.

For the ducks to sit on and enjoy the sun, think about including logs or rocks. These elements provide an environment that is varied and promotes play and exploration.

Pond Upkeep And Safety Advice

To keep your ducks safe and healthy, you must regularly maintain your duck pond. Here are some vital upkeep pointers:

Frequent Cleaning

To get rid of waste, extra feed, and debris, clean the pond frequently. Waste buildup might result in low water quality and health problems for your ducks. Effective maintenance techniques include using a siphon to drain and clean the bottom and a skimmer net to remove floating debris.

Levels of Water

Keep an eye on the pond's water level, particularly when there are dry spells or a lot of rain. For ducks, maintaining a steady water level is essential since variations can stress them and cause health issues. Consider adding a water source, such as a pump or hose, to refill your tiny pond.

Control of Predators

It's critical to keep your ducks safe from possible predators. In addition to keeping a watch out for regular predators like foxes, raccoons, and birds of prey, think about erecting fencing around the pond area. Adding hiding places to the pond, like thick bushes or man-made shelters, can also make ducks feel safer.

Considering the Seasons

Pond upkeep may be impacted by seasonal variations. Make sure the pond doesn't freeze over completely throughout the winter. Even in cold weather, ducks can still gain from having access to water. To keep a small area of open water, you may need to drill a hole in the ice or use a pond heater.

Safety Procedures

When maintaining your ducks' water source, always put their safety first. Refrain from utilizing any dangerous materials or chemicals close to the pond as they may contaminate the water. Keep an eye out for any signs of concern in your ducks' behavior and regularly check for any sharp items or trash that could cause them damage.

In summary

Slaughtering ducks requires careful attention to water and pond management. You can make sure your ducks have a healthy and stimulating habitat by appreciating the value of water, designing a suitable pond, and putting good management practices into place. Cheers to raising ducks!

CHAPTER NINE

Raising Ducks For Their Eggs And Meat

Whether you want to raise ducks for meat, eggs, or both, duck farming can be a lucrative endeavor. As a general rule, ducks are harder to handle for novices than chickens. They can be a wonderful addition to your farm or backyard and have clear advantages in terms of meat quality and egg production. The fundamentals of raising ducks for meat and eggs will be covered in this section, with an emphasis on the requirements unique to each product.

Growing Ducks For The Production Of Meat

Comprehending the many kinds of ducks is crucial while rearing them for meat production. Some breeds are more adapted for meat than others; one of the most widely used varieties is the Pekin duck. Pekins can be plucked as early as seven or eight weeks and are prized for their soft meat and quick growth. Some breeds, like the Khaki Campbell and the Muscovy, are equally excellent at producing meat, but they could take longer to reach market weight.

Selecting the Appropriate Breed

Investigate the breeds that will produce the most for your climate and farming objectives before you begin. Growth rate, feed conversion ratio, and meat quality are important factors to take into account. After selecting your breed, make sure you have the necessary spaces set up to foster their development.

Environment and Housing

For them to live well, ducks need a secure and pleasant environment. Their home should offer them the freedom to roam, proper ventilation, and defense against predators. All you need to protect them from bad weather is a basic duck home with enough insulation. Water is another resource that ducks like to have available for drinking and foraging. They'll stay content and healthy with a little pond or kiddie pool.

Nutrition and Feeding

For the creation of meat, proper nutrition is essential. Ducks require a well-balanced diet consisting of premium feed that is high in protein and other vital components. For ducklings, starter meals are usually advised; as they become older,

they switch to grower foods. Observe their feeding schedule and make sure they always have access to clean water. To improve their growth and flavor, it's also advantageous to add kitchen scraps or foraged greens to their diet.

Gathering And Preparing Ducks

When your ducks get to the weight you want, you should think about harvesting them. To maintain food safety when processing ducks, meticulous planning and adherence to hygienic standards are necessary.

When to Harvest

Depending on the breed and your production objectives, there are different times of year to harvest ducks.

Most meat ducks can be harvested between the ages of 7 and 12 weeks. To determine the ideal time for processing, keep a close eye on their weight and condition.

Methods of Processing

Ducks can be processed in a variety of ways, from small-scale, manual processes to more sophisticated machinery for bigger enterprises. When handling a limited quantity of ducks, manual plucking might be adequate. To save time and labour when working with larger batches, think about purchasing a plucker.

Considering Food Safety

It's critical to keep everything tidy during the processing procedure. To avoid contamination, make sure that all surfaces and equipment are thoroughly cleaned. To guarantee that poultry products are processed and sold in accordance with local laws, observe health rules.

Gathering and Storing Eggs

Ducks can lay a lot of eggs, and because of their excellent nutritious content and delicious flavor, people prize them much. It is essential to know how to handle egg collection and storage if you want to maximize your output.

Recognizing Duck Breeds To Produce Eggs

While different varieties lay different quantities of eggs annually, one of the best producers is the Khaki Campbell, which may lay up to 300 eggs. Some breeds, like the Indian Runner, also lay a significant quantity of eggs. To make sure your expectations are in line, find out whether the breed you have picked can lay eggs.

Procedures for Collecting Eggs

To keep eggs clean and free from harm, it's imperative to collect them on a regular basis. When egg-laying is at its best, try to gather eggs twice a day, at the very least. To facilitate egg collection, provide clean bedding in nesting boxes to encourage ducks to lay their eggs in certain locations.

Appropriate Strategies for Storing Eggs

To preserve freshness, eggs must be stored correctly after collection. To extend their shelf life, store eggs in a cool, dry location—ideally in the refrigerator. You may monitor the eggs' freshness by marking the collection date on the label.

promoting duck eggs

After you've got a reliable system in place for producing eggs, you might want to market your duck eggs. Consumers may find duck eggs more enticing than chicken eggs because they have richer yolks and thicker shells. To sell your eggs, look into regional farmer's markets, supermarkets, and internet marketplaces. To draw in customers, emphasise their special features, such as their nutritional advantages and culinary applications.

CHAPTER TEN

Trader Of Meat, Eggs, And Ducks

Recognizing Your Products

It's critical to comprehend the kinds of things you'll be marketing before entering the industry. Ducks are prized for their eggs as well as their meat. Different consumer bases may find different duck breeds, such as Pekin or Muscovy, appealing when it comes to meat. In a similar vein, duck eggs are highly prized for their flavour and culinary applications. Effective marketing will start with knowing what makes your ducks, their meat, and their eggs special.

Duck Product Types

1.Live Ducks: Whether you're selling pets, show ducks, or home flock additions, you can draw in customers by offering live ducks.

2. Duck flesh: Premium duck flesh can be frozen, marketed fresh, or made into specialty products like confit or sausage.

3. Duck Eggs: Preferred by gourmet chefs and health-conscious consumers, duck eggs are richer and larger than chicken eggs.

4. Duck down and feathers are byproducts that can be sold for use in clothes and bedding, opening up new revenue streams.

Identifying Your Market
Finding Possible Clients

The first step in making successful sales is determining your target market. Customers who might buy duck items include:

• Local Restaurants: A lot of them look for unusual foods. Speak with chefs who specialize in farm-to-table or gourmet cuisine.

• Farmers' Markets: Having a booth at a nearby farmers' market enables you to speak with customers face-to-face and display your goods.

• Grocery Stores: Locally sourced duck items can be of interest to specialty shops and smaller grocery stores.

• Direct-to-Consumer Sales: You can contact customers directly by using social media, your website, or neighborhood associations.

Recognising Customer Preferences

It's important to know what kinds of customers you might attract. Find out what aspects of local food consumers' diets, like organic or free-range items, they value. Use social media polls and surveys to Interact wIth your audience and learn about the things they are interested in buying.

Developing Connections

Developing a solid rapport with your clients can boost their loyalty and promote repeat business. At neighborhood gatherings, provide samples, interact with clients online, and pay attention to their comments. Having a strong brand presence might make you stand out from the competition.

Product Packaging and Sales for Ducks
The Value of Packaging

Packaging is essential to the safety and appeal of a product. Superior packaging conveys the values of your company while simultaneously safeguarding your products. Regarding duck-related products, take into account:

• Sustainability: Consumers who care about the environment may be drawn to eco-friendly packaging. When feasible, make use of recyclable materials.

• Transparency: Provide information on the product's origins, farming methods, and nutritional facts on its labels.

Channels of Sale

1.Online Sales: You can reach a wider audience by opening an online store. Direct sales can be facilitated by websites like Etsy, Shopify, or even local delivery applications.

2. Farm Stands: Establish a farm stand on your land. Customers can visit as a result, discover more about your farming methods, and buy fresh goods.

3. Wholesale: Consistent sales can be achieved by entering into wholesale relationships with nearby restaurants or grocery stores.

Making Eye-Catchy Displays

An eye-catching display might attract clients to your farm stand or at markets where you trade. Promote the qualities and advantages of your products with signage, and think about offering samples to get customers to try them out. A neatly set up exhibit makes a good first impression and may boost sales.

Beginner's Guide To Pricing Strategies
Performing Market Research

Do a comprehensive market analysis to determine the going prices for duck items in your area before deciding on a price. Talk to other farmers, visit nearby markets, and look up prices online. This study will give your pricing approach a starting point.

Analyzing Costs

Make sure your prices cover expenses by calculating your production costs. Add:

• Feed Costs: Keep in mind that ducks need a balanced diet, so account for this.

• Housing and Equipment: Take into account the price of your water supply, duck housing, and additional equipment.

• Labour Costs: Make sure your pricing accounts for labor costs if you employ assistance.

Models of Pricing

1. Cost-Plus Pricing: Increase the cost of production by a margin. This simple strategy guarantees you break even and turn a profit.

2. Competitive Pricing: Base your rates on those of your rivals. While maintaining your profit margin, this might help you stay competitive.

3. Value-Based Pricing: Take into account setting prices depending on perceived value if you sell high-end goods (like organic or free-range ducks). Consumers could be prepared to spend extra for superior or distinctive features.

Providing Sales and Marketing

Providing introductory deals or packages can draw in new clients and promote recurring business. In order to keep customers interested, think about loyalty programs or seasonal specials.

Keeping an eye on and modifying prices

Keep an eye on market trends and your sales at all times. Be ready to modify your prices in response to changes in production costs, feedback, and demand. Being adaptable will help you keep your edge in the competition.

CHAPTER ELEVEN

Planning Finances And Managing Farms

Successful duck farming requires careful planning, especially about finances and farm management. Comprehending the financial components as a novice will aid in making well-informed judgments, guaranteeing the longevity and financial success of your farm. The fundamentals of financial planning, continuing expense management, and growth planning will all be covered in this part.

The Price Of Beginning A Duck Farm
First Invested

Understanding the early expenses involved in starting a duck farm is essential. Usually, the biggest costs are related to building a proper home, buying ducklings, and acquiring the required equipment. Ducklings vary in cost based on the breed but expect to pay between

$3 and $10 per bird. Furthermore, think about spending $500 to several thousand dollars, depending on the size and materials chosen, on a secure and robust duck home.

Terrain and Facilities

One of your biggest investments, if you don't already own land, will be buying or renting suitable property. For swimming and foraging, ducks need access to clean water, so look for areas with plenty of water sources. Land prices might vary greatly depending on where you live, so do your homework on local real estate values. To provide your ducks with a secure and productive environment, you might also need to make investments in feeders, water troughs, and fencing.

Food and Materials

You should account for the recurring expense of feeding your ducks when creating your initial budget. A good diet is necessary for the growth and productivity of ducks. You should budget between $15 and $30 for each bag, depending on the recipe and quality of the feed. Think about the expenses of vitamins, bedding, and medical supplies in addition to feed, as these can quickly mount up.

Handling Recurring Charges
Water and Food

The majority of your continuing expenses will probably come from feeding, which is a recurrent expense. A balanced diet that includes grains, commercial feed, and leftover food from the kitchen is necessary for ducks. Make sure you factor in the number of ducks and their developmental phases when calculating the monthly feed expenses. Another important component is water; make sure there is always enough available for swimming and drinking. Inadequate water access can have an impact on the productivity and health of ducks.

Veterinary and Medical Expenses

Sustaining the well-being of your flock is critical to output and financial gain. Medication, vaccines, and routine veterinary examinations can be very expensive. Set aside money for regular health examinations and prepare for any medical emergencies. For your ducks to receive the care they need, it's a good idea to build a relationship with a veterinarian who specializes in poultry.

Labour Charges

You might require hiring more staff if your business is bigger. Farmhands for general maintenance, cleaning, and feeding could fall under this category. Salary ranges might change depending on the labor market and where you live. Think over whether you'll be managing the farm by yourself or if you'll need to budget for employing staff.

Developing A Growth And Profitability Plan

Establishing Objectives and Goals

Setting definite goals and objectives is crucial when starting a duck farming operation. Choose if you want to grow meat, eggs, or both. Establishing quantifiable goals for output will enable you to monitor your progress and make the required corrections. Having a clear vision will direct your financial planning, regardless of your goals—growing your business or branching out into value-added items.

Market Analysis

Comprehending your intended audience is essential for financial success. Find possible clients by conducting in-depth market research at places like neighborhood eateries, supermarkets, and farmers' markets. Examine local duck product demand and pricing patterns. Your production choices and marketing plans will be informed by this data, which will help you match your product offers to the demands of the market.

Investing in Technology and Equipment

As your duck farm expands, think about making an investment in cutting-edge machinery and technology to boost production and efficiency. Labor costs can be decreased and operations can be streamlined with the use of automated climate control systems, waterers, and feeding systems. Even while these upfront costs could be high, they can eventually pay for themselves with lower manual labor costs and higher production rates.

Maintaining Financial Records

Keeping correct financial records is essential to managing a farm successfully. Maintain a record of all earnings and outlays, such as labor costs, feed costs, and veterinary expenditures. Spreadsheets or accounting software can help you keep your funds organized. Examine your financial statements regularly to determine your profitability and to help you decide whether to scale or modify your business.

CONCLUSION

For novices, duck farming is a very fulfilling endeavor that combines the love of animal husbandry with sustainable agricultural methods. We have looked at many facets of duck farming in this book, from identifying the various varieties to creating an atmosphere that is conducive to their development. It's time to advance in your duck farming endeavors now that you possess the necessary foundational information.

Seize the Opportunity to Learn

Newcomers need to understand that there is a learning curve associated with duck farming, just like with any other farming endeavor. Every flock will provide different chances and challenges. Accept these times as a necessary part of the adventure, whether you're handling health concerns, figuring out what they should eat, or controlling their breeding cycles. Join local agricultural clubs, social media groups, or forums to interact with other duck farmers and exchange experiences and knowledge.

Developing Ideal Procedures

The success of your farm and the welfare of your ducks depend on the establishment of best practices. Keep a close eye on their health, feed them a healthy food, and keep their living space tidy. With more experience, you'll be able to spot such issues before they become serious, keeping your ducks healthy and productive. Putting best practices into effect will also improve the meat and eggs from your ducks, adding value to your farming endeavors.

Ethics and Sustainability

Another sustainable business that benefits regional ecosystems is duck farming. Ducks are renowned for their foraging skills, which enhance soil health and aid in natural pest control. It's imperative to use ethical farming methods; make sure your ducks' needs are satisfied by treating them with kindness and respect. In addition to advancing animal welfare, a humane farming method can draw in consumers who appreciate goods with ethical origins.

monetary considerations

Take the financial aspects of duck farming into consideration before you get started. Create a budget that accounts for initial outlay, continuing costs, and anticipated revenue from the sale of meat or eggs. It's critical to keep close tabs on your finances and modify your procedures in response to changes in the market and operating expenses. To assist them in making well-informed decisions about growing their businesses or broadening their product offers, many prosperous farmers keep thorough records.

Last Words

As you start your duck farming endeavor, keep in mind that perseverance and patience are essential. You may create a successful duck farm that offers financial gains as well as personal fulfillment with hard work and the appropriate strategy. Follow your passion, educate yourself, and strive to make your methods better at all times. Happy agribusiness!

THE END

www.ingramcontent.com/pod-product-compliance
Lightning Source LLC
Chambersburg PA
CBHW052330220526
45472CB00001B/355